THE POWER OF THE SUN

The Sun and the Weather

Alicia Z. Klepeis

Cavendish Square
New York

Published in 2020 by Cavendish Square Publishing, LLC
243 5th Avenue, Suite 136, New York, NY 10016

First Edition

Library of Congress Cataloging-in-Publication Data

Names: Klepeis, Alicia, 1971- author.
Title: The sun and the weather / Alicia Z. Klepeis.
Description: First edition. | New York : Cavendish Square, [2020] | Series: The power of the sun | Audience: Grades 2 to 5. | Includes bibliographical references and index.
Identifiers: LCCN 2018047194 (print) | LCCN 2018048274 (ebook) | ISBN 9781502646736 (ebook) | ISBN 9781502646729 (library bound) | ISBN 9781502646705 (pbk.) | ISBN 9781502646712 (6 pack)
Subjects: LCSH: Weather--Effect of solar activity on--Juvenile literature. | Atmosphere--Juvenile literature. | Sun--Juvenile literature.
Classification: LCC QC883.2.S6 (ebook) | LCC QC883.2.S6 K54 2020 (print) | DDC 551.5--dc23
LC record available at https://lccn.loc.gov/2018047194

Editorial Director: David McNamara
Editor: Jodyanne Benson
Copy Editor: Nathan Heidelberger
Associate Art Director: Alan Sliwinski
Designer: Jessica Nevins
Production Coordinator: Karol Szymczuk
Photo Research: J8 Media

The photographs in this book are used by permission and through the courtesy of: Cover Underworld/Shutterstock.com; p.3 (used throughout) Black Prometheus/Shutterstock.com; p. 4 Leedsn/Shutterstock.com; p. 6 Soleil Nordic/Shutterstock.com; p. 7 Frame 18 Productions/Shutterstock.com, (used throughout) St-n1ce/Shutterstock.com; p. 8 Rvector/Shutterstock.com; p. 10 E Forafontova/Shutterstock.com; p. 11 Karsten Schneider/Science Source; p. 12 Nasky/Shutterstock.com; p. 13 Brgfx/Shutterstock.com; p. 14 Gary Hincks/Science Source; p. 16 Honza Krej/Shutterstock.com; p. 18 Vitoriano Junior/Shutterstock.com; p. 20 Soleil Nordic/Shutterstock.com; p. 21 Soleil Nordic/Shutterstock.com; p. 22 Tim Brown/Science Source; p. 23 Claude Huot/Shutterstock.com; p. 24 Anna Frajtova/Shutterstock.com; p. 26 Pressmaster/ Shutterstock.com; p. 27 Simon Dannhauer/ Shutterstock.com; p. 28 Sergey Novikov/ Shutterstock.com.

Printed in the United States of America

Contents

Rainbows form when there are water droplets in the air and sunlight shines through them.

Chapter 1

The Sun Creates Weather

LOOK out your window. It might be sunny or rainy. It might be windy or calm. Perhaps it's hot out. Maybe it's super chilly. All of these words describe the weather. But what causes weather? What creates the different conditions we find when we step outdoors? The sun!

The Sun's Energy

The sun is a powerful source of energy. It also is the main reason that Earth

The sun's rays travel millions of miles to reach Earth's surface.

has weather. Each day, the sun sends out, or radiates, a huge amount of energy. Some of it travels all the way to Earth. That's a distance of about 93 million miles (150 million kilometers)!

Just like the sun warms your skin on a hot summer day, it also warms the surface of Earth. The energy of the sun's rays heats up the air, the land, and the oceans all over our planet. This is what creates our weather.

One day it might be warm enough for you to wear shorts and a T-shirt to school. Then, the wind blows overnight while you are sleeping. You might need pants and a sweater the next day. Weather is always changing!

SUN POWER

The average surface temperature on Earth is about 57 degrees Fahrenheit (13.9 degrees Celsius). The surface temperature of the sun is about 9,932 degrees Fahrenheit (5,500 degrees Celsius)!

Weather and Dances

Weather plays an important part in people's daily lives. Around the world, different groups of people celebrate the sun's importance through dance. Many Native American tribes from the Great Plains have a sun dance ceremony in late spring or early summer. Dancers paint their bodies with different colors. They also make music with bells and drums. Rain dances are also common in **indigenous** cultures from Australia to the United States. People hope that by dancing, the rain that they need to survive will come.

SUN POWER
The sun produces more energy in one second than humans have used in history.

Colorful clothing and music are part of this Native American rain dance ceremony.

These colorful pictures show many weather-related items and events, from tornadoes to rainstorms.

Chapter 2

Earth's Atmosphere

WEATHER is the state of Earth's atmosphere. The atmosphere is a huge mass of air surrounding Earth. It's made of gases. These gases make up the air we breathe every day.

When sunlight reaches Earth, it heats up the atmosphere. Our planet's atmosphere absorbs much of the sun's energy. It keeps Earth's surface from becoming too cold or too hot.

Day and Night

Every twenty-four hours, Earth spins around once on its **axis**. This is the reason we have day and night. During the day, sunlight heats up the part of the world facing the sun. Temperatures rise as these places absorb the sun's heat.

As Earth rotates on its axis, part of the world experiences daytime while other areas have nighttime.

When the part of the world where you live is turned away from the sun, it's nighttime. During the dark hours, Earth's surface cools. Places gradually lose the heat they absorbed over the course of the day. Temperatures drop.

Different Places, Different Weather

Because Earth is shaped like a ball, sunlight hits different places on Earth at different angles. So the sun heats some locations more than others. The warmest places in the world are found along the **equator**. Here, the sun is nearly overhead at noon. Strong rays of sunlight hit this area, making temperatures warm or hot.

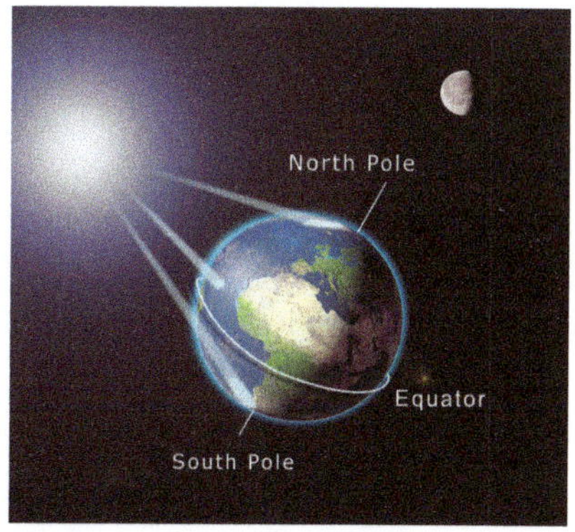

This diagram shows how the area near the equator receives more direct sunlight than areas near the poles.

The world's coldest places are near Earth's poles. The sun's rays here tend to be spread out over a much wider area than near the equator. Without strong direct sunshine, temperatures stay cold.

Between the equator and each pole is a temperate zone. Weather here can vary from hot to cold, depending on the strength of the sunlight hitting this area. The sun's rays are generally weaker as you move away from the equator. The higher **latitudes** receive less of the sun's heat.

SUN POWER
The fastest wind speed recorded on Earth was 253 miles (407 kilometers) per hour!

Winds and Pressure Systems

Sunlight also makes the air around you calm or windy. The uneven heating of Earth's surface by the sun causes wind. When the sun warms up the air in a particular place, that warm air spreads out and rises. This creates what's called a low-pressure area.

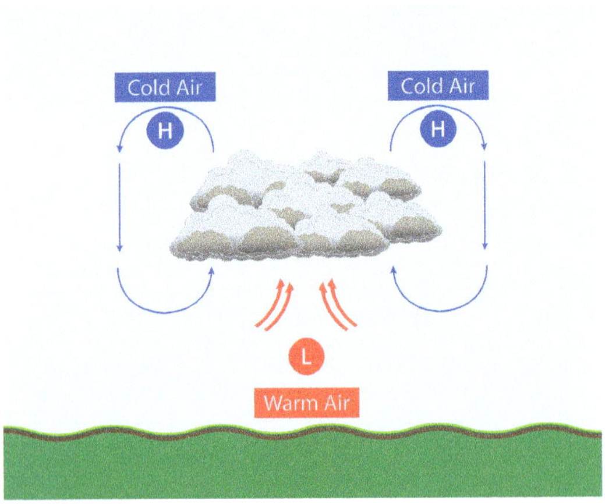

Low pressure areas are where warm air rises. High pressure areas are where cold air sinks.

Cooler and heavier air flows in from surrounding

areas and takes the place of the warm air. An area where cooler air is sinking is known as a high-pressure zone. Air flows from areas with high pressure to those with low pressure, creating winds. If Earth's surface was all the same temperature, there wouldn't be winds.

SUN POWER
Antarctica is the windiest place on Earth, with winds that commonly reach 200 miles (322 km) per hour!

What Falls from the Skies

The sun is an important part of why **precipitation** happens. Heat from the sun causes both liquid and frozen water to evaporate into a gas called water vapor. This gas rises way up in the sky to form clouds. These clouds travel around the world. Sometimes they drop rain. Other times they drop snow.

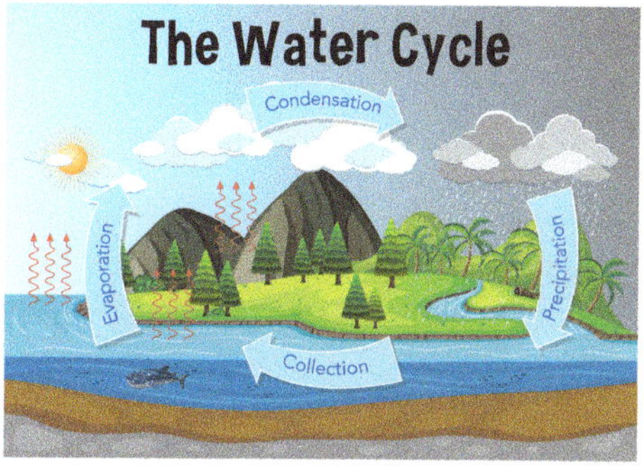

Water moves between the land, the atmosphere, and the oceans and rivers in the water cycle.

Global Wind Patterns

A global system of winds moves air around the world. Since the sun's direct rays hit the equator, the air here gets heated and rises. It moves toward the poles. This warm air starts to cool and sink around 30 degrees latitude. Most of this sinking, cooling air flows back toward the equator. These air movements toward the equator are known as the trade winds.

There are also winds that move toward the poles. Between 30 and 60 degrees latitude, the winds are called prevailing westerlies. These winds are responsible for many weather movements in the United States and Canada. Around 60 degrees latitude, the polar easterlies form. Here, cool air sinks and flows away from the poles.

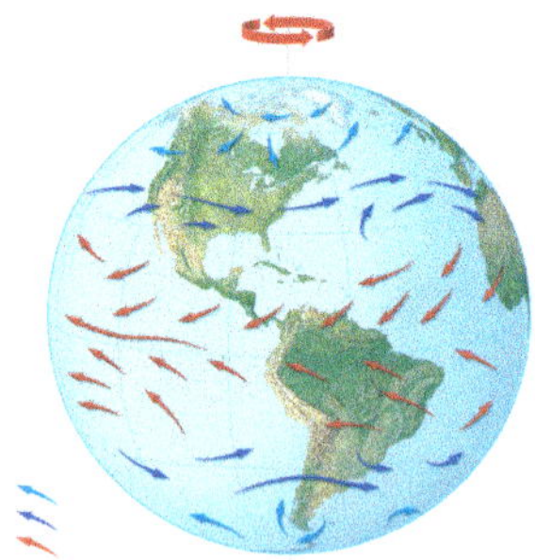

This model shows the wind patterns found in Earth's atmosphere. The trade winds are shown in red.

ACTIVITY: Make a Rain Gauge

Try making this simple rain gauge. How much precipitation falls where you live?

Supplies:

- Empty 2-liter plastic bottle
- Scissors
- Ruler
- Permanent marker
- Duct tape

Directions:

— Cut about 4–6 inches off the top of your bottle.

— Put this bottle top, upside down, inside the other piece (bottom) of the bottle.

— Tape around the top of where the two bottles connect to keep them in place.

— Using your ruler and permanent marker, measure and mark in 0.25-inch increments from the bottom of the bottle until 4 inches.

— Put your bottle on the ground.

— Each day, record how much rain or snow falls into your bottle. Be sure to empty the bottle after measuring.

Spring flowers, colored autumn leaves, and winter snow are examples of the seasonal changes we experience each year.

Chapter 3

Reasons for the Seasons

EVEN though the sun shines down on Earth all year long, the seasons change. There are four seasons: spring, summer, autumn, winter. Each has its own particular patterns of weather and daylight hours. Why are there seasons? Are the seasons the same everywhere on Earth?

The Earth Is Always Moving

Before you can understand seasons, it's important to know about Earth's movement. There are two ways Earth moves all

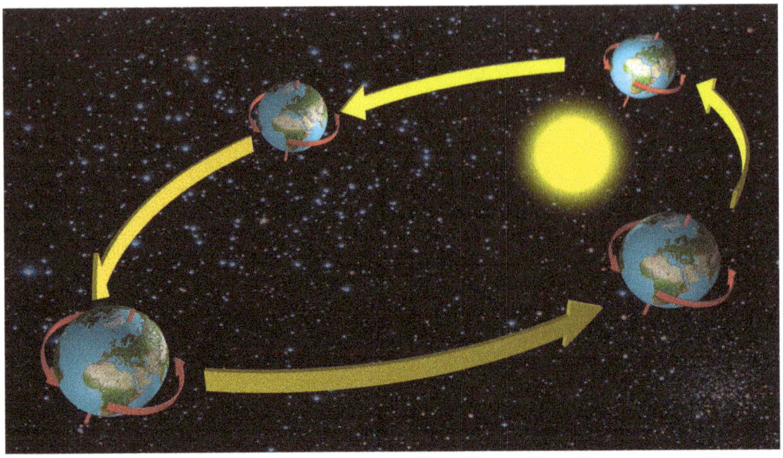

Red arrows show Earth rotating on its axis. Yellow arrows show Earth revolving around the sun.

the time. One is rotation, which is when Earth spins on its axis. The other movement is called revolution. This is when Earth travels around the sun. It takes about 365 days, or one year, for Earth to complete one orbit around the sun.

Earth and Its Seasons

There are two reasons why we have seasons. One is Earth's axis. The other has to do with Earth's revolution.

If you've ever seen a globe, you know that Earth doesn't stand straight up and down.

It's tilted. Because Earth is at an angle, one of its **hemispheres** leans toward the sun for part of each year. As Earth travels around the sun, the section of Earth's surface that leans toward the sun changes.

Sun in the Summer

When a hemisphere is leaning toward the sun, it receives a lot of direct sunlight. The sun's rays are hitting that part of Earth straight on. It's summertime. The sun is high in the sky during summer. The long daylight hours give that part of Earth lots of time for the temperatures to get warm.

Sun in the Winter

When a hemisphere is tilted away from the sun, it gets indirect sunlight. The sun's rays are more spread out.

SUN POWER
During the winter in Longyearbyen, Norway, it stays dark for about four months.

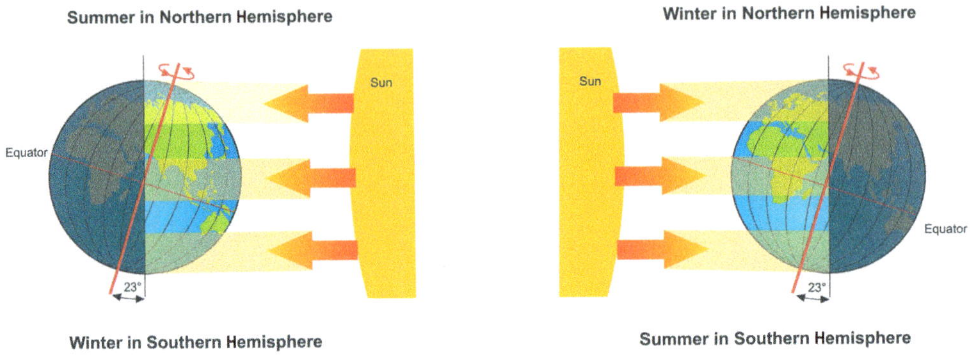

When a hemisphere is tilted toward the sun, it is summer there. When that hemisphere is tilted away from the sun, it is winter.

It's wintertime. The sun doesn't get that high in the sky during winter. Less energy hits this part of Earth's surface. The long nights and short days keep Earth's surface from getting very warm. Temperatures are cooler.

Spring and Autumn Sun

In spring and autumn, neither hemisphere is tilting away from or toward the sun. Everywhere on Earth receives about the same amount of direct sunlight. Temperatures are more moderate. They are not too hot or too chilly. There's much less difference between the length of the days and nights.

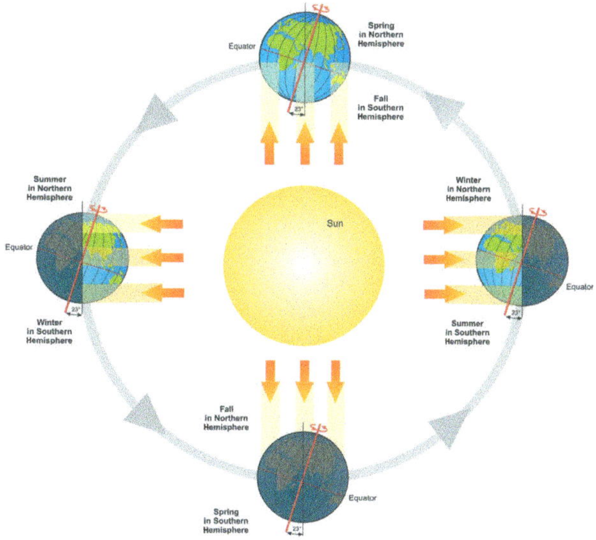

This illustration shows how the sun's rays hit Earth during all four seasons.

Opposite Seasons or No Seasons

Seasons in the Northern and Southern Hemispheres are opposite. When it's summer in the United States, it's winter in Australia.

Some places don't have big temperature changes between seasons. One example is near the equator. Why? The sun is always high in the sky here. It's warm year-

SUN POWER

On the spring and autumnal **equinoxes**, day and night are about the same length of time everywhere on Earth.

Why Hurricanes Are Seasonal

Hurricanes are huge storms with very strong winds. Hurricanes need warm ocean water (over 80°F, or 27°C) to form. The Atlantic hurricane season lasts from June 1 to November 30. However, most storms happen in the fall. That's because the ocean waters are warmed up by the sun's rays over the course of the Northern Hemisphere's summer, leading to lots of hurricanes when fall comes around.

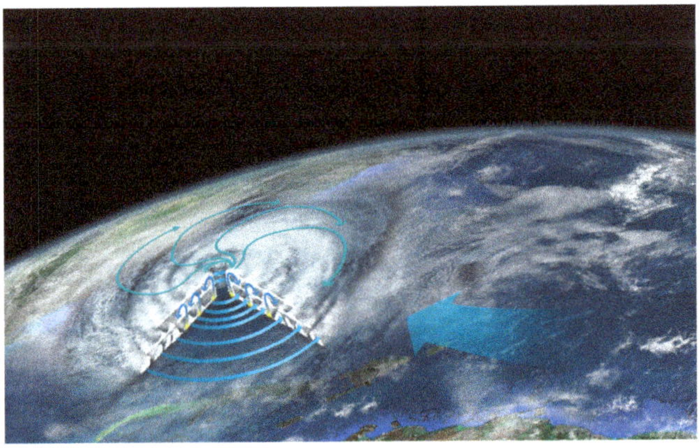

Hurricanes form over the ocean's surface. This image shows the central point around which winds circulate (spin).

round. In the tropics, it's common to just have two seasons: wet and dry. Temperatures don't vary much throughout the year here. However, the temperate zones have distinct weather patterns in the four seasons.

This photo shows Africa's Victoria Falls in the dry season. Much more water flows over the falls in the wet season.

ACTIVITY: What Would You Wear?

With the help of a parent, teacher, or librarian, research the average temperature of where you live in January and July. Then, find out what the weather would be like in a city located in the opposite hemisphere during those same months. So, if you live the Northern Hemisphere, you might choose to research the weather in Sydney, Australia, because it is located in the Southern Hemisphere. Finally, draw what type of clothes someone would wear in each location, both in January and in July.

Many people check the weather forecasts on their mobile phones.

Chapter 4

No Sun, No Weather

THE sun is a powerful source of energy. It's also the main reason that Earth has weather. Every day, sunlight travels millions of miles from its source to where we live. This light warms up the land and oceans all over the world. It also heats up the gases that make up Earth's atmosphere.

Sunshine Is Life

Without the sun, Earth would be impossible for humans to live on. People need the sun's

Children play outside on an overcast day. Their clothes suggest the temperature is warm.

energy to make temperatures comfortable on our planet. Not everywhere on Earth has the same weather. Some places are hotter or colder than others. That's because the sun does not evenly heat the whole planet. Locations near the equator receive more

SUN POWER
While the sun is highest in the sky at noon, 3 p.m. is the hottest time of day.

A boardwalk runs through the swamp in Florida's Everglades National Park. It's hot all year round there.

direct sunlight than those near the poles, leading to warmer weather overall.

Seasons and Sunlight

When you think of weather, the seasons probably come to mind. The tilt of Earth's axis and the planet's orbit around the sun

In the winter, kids play in the snow. They are dressed for cold weather.

are what cause us to have seasons. As Earth revolves around the sun, one hemisphere tilts toward the sun. It receives more direct sunlight and more of the sun's heat. This is summertime. When a hemisphere is tilted away from the sun, it receives indirect sunlight and less of the sun's heat. This is wintertime.

Without the sun, we wouldn't experience different kinds of weather. One thing's for sure—the sun rules our weather!

SUN POWER
Volcanic eruptions can block the sun's rays and make weather cooler for years.

Glossary

axis The imaginary line around which Earth rotates.

equator An imaginary circle around the center of Earth, equally distant from the two poles.

equinox When the sun appears overhead at the equator, and day and night are the same length.

hemisphere One half of Earth, as divided by the equator.

indigenous Native to a particular area or environment.

latitude An imaginary circle around Earth marking the distance north or south of the equator.

precipitation Water that falls to Earth as rain, snow, mist, sleet, or hail.

Find Out More

Books

Gibbons, Gail. *The Reasons for Seasons.* New York: Holiday House, 2019.

Rattini, Kristin Baird. *Weather.* Washington, DC: National Geographic Kids, 2018.

Websites

DK Find Out!: What Causes Weather?

https://www.dkfindout.com/us/earth/weather/what-causes-weather

Interactive graphics on this website explain how the sun causes weather.

PBS Learning Media: Clouds and Weather

https://www.pbslearningmedia.org/resource/nvcl.sci.earth.weather/clouds-and-weather

Explore the relationship between the sun, weather, and clouds with this interactive website.

Index

Page numbers in **boldface** refer to images. Entries in **boldface** are glossary terms.

About the Author

Alicia Klepeis began her career at the National Geographic Society. She is the author of more than ninety children's books, including *The World's Strangest Foods*, *How Smartphones Work*, and *A Time for Change*. Klepeis lives with her family in upstate New York. Her favorite season is autumn, with its sunny, crisp days.